Wazzat Beetle?

A Guide to Beetles of Culpeper County and Northern Virginia

Lee Alloway

Ancient Eagle Press
Culpeper, Virginia

ISBN: 978-0-9971923-3-9
Library of Congress Control Number: 2022900913

To the world's great recyclers: The Beetles.

PROLOGUE

I never intended to write about beetles. Airplanes, sure. I know about airplanes. Travel? I've done plenty and can probably offer a few insights. But beetles? They were nowhere on my radar. They were neither mysterious nor interesting. Lady bugs were pretty and I knew they ate aphids, but beyond that beetles were like air and sunlight. I knew they existed and didn't care to pursue the issue any further. Then I retired. Came up for air. Started looking around. With camera in hand, I started inspecting my yard for small things. Flowers at first, then birds, spiders, mushrooms. Then came the bugs. The more I looked, the more I noticed. I was amazed to find fifty different species within a few hundred feet of the house. Fifty became a hundred, then two hundred, then a thousand species. Moths, butterflies, spiders, true bugs, and lots of beetles. Beetles in many different shapes, sizes, and colors, existing in every sort of environment from the lake out back to the hardscaping in the driest part of the yard. Some in trees, some underground. Some were eating our garden; some were eating each other. Beetles, as it turns out, are a diverse and interesting group, and well worth getting to know. Since then, I've traveled around the world photographing bugs, but our own home-grown beetles have a special place in my yard and in my heart.

Contents

Introduction

Beetles! They're everywhere, and no wonder. Of the roughly one million bug species identified by scientists, about 400,000 of them are beetles. Given that there are an estimated ten quintillion bugs in the world, the total beetle population is certainly astounding...and that's fortunate. Without beetles, life as we know it would not exist.

- Without beetles, we couldn't stop and smell the roses because we would be hip deep in dead things. Beetles are the front line workers recycling deceased animals, trees, and cow patties.
- Without beetles a day at the lake would be miserable, since it is aquatic beetles that help keep the mosquito population in check.
- Without beetles the entire food chain would work differently, since beetles are fundamental prey and/or predators of animals throughout the food chain.
- Without beetles, humans might have evolved differently, as beetles have been a source of fat and protein in the diet of humans since the beginning of humanity.

The importance of beetles is reflected by their place in religion and folklore, folk medicine, food, and ornamentation. The adoration of the scarab *Scarabaeus sacer* (a dung beetle) in Egypt is well known[1], but other scarabs have a place in the cosmology of many early European cultures as well. African, European, and Asian cultures have all embraced beetles in their folk medicines.

Beetles have been and in many parts of the world continue to be a common food. Entomophagy is on the decline in Europe and has never been particularly popular in the United States, but into the 20th Century insects continued to be a delicacy in European cuisine. Numerous species

---

[1] The earliest example in ancient Egyptian iconography of a scarab was of another scarab, *Kheper aegyptiorum*. The god Khepri is depicted as having the head of a scarab. Over several thousand years of the Egyptian dynasties, several different scarab species were favored.

– particularly scarabs, thanks to their relatively large size – are still eaten throughout Asia, Africa, and South America. Since this book is about beetles of Culpeper County, it should be noted that one place that does not have a documented history of eating beetles is North America. Among the indigenous populations, insects in general show up occasionally in the folk lore or medicine but were rarely on the menu. This may be due to the relative abundance of other sources of protein.

Of the 400,000+ described beetle species in the world, an estimated 15,000 species are found in the Eastern United States. There are doubtless many more that haven't been identified yet. Beetles are everywhere, but if you live in the city, you may be hard pressed to find them. They tend to be small and not fond of concrete and asphalt. Head to the park or the suburbs, and you will be astounded at what you can find. Some may look a bizarre or scary, but with few exceptions beetles are harmless to humans.[2]

With 15,000 Eastern North American beetle species to choose from, how do you know what you are seeing when you find one? Grab a plastic cup, capture the invader, and spend some time studying it. Look at the colors, the amazing eyes, the complex legs. If you have a field guide handy, you will find hundreds of bug pictures, many of which look a bit like your bug, none of which look exactly like it. If you turn to the Internet, your choices are almost limitless, offering hours of entertainment before you discover the bug you thought looks most like yours happens to live only in the Andes above 14,000 feet. But with *Wazzat Beetle?* you have a fighting chance, since this book contains pictures only of beetles found in Northern Virginia[3]. So look around, grab a plastic cup, go find a bug and ID the critter before you let it go about its business.

---

[2] Rove beetles of the subfamily Paederinae contain pederin, which causes contact dermatitis if it is rubbed against bare skin. See discussion on page 56-57.

[3] All the bugs in this book were photographed in Culpeper, Orange, or Fairfax counties. All were photographed alive where observed. Some were additionally photographed in a white box before being returned to their natural environment.

Definitions

First let's agree on some definitions:

Bug: For purposes of this guide, a "bug" will be any small creature with six or eight legs in its adult form.

Insect: A bug that has

- Three pairs of legs
- Three body parts: head, thorax, and abdomen
- Typically, two pair of wings[4];
- One pair of antennae
- Usually, one pair of compound eyes

Almost one million species of insects have been identified, with millions more still to be described. Beetles alone count for almost 400,000 described species, with an estimated one million more species crawling around undescribed.

Beetle: An insect in which the forewings have hardened into protective coverings called *elytra* (singular: *elytron*).

Harmful/beneficial: Bugs are not inherently harmful or beneficial. They are simply doing what they are designed to do. For the most part, we don't see the bugs around us and we have no opinion one way or the other about "harmful" or "beneficial." It is only when the normal function of the bug intersects our own lives that we pass judgement. If we have a vegetable garden we may be competing with "harmful" insects for crops; raising flowers, we may find insect damage reduces the commercial or esthetic value of our product. A beetle may be "beneficial" when it kills caterpillars that eat your flowers, but "harmful" when it kills your squash vine.

---

[4] In some families one pair of wings has evolved into something else. In diptera (flies) one pair of wings has become a stabilizing structure called a haltere. In beetles, the forewing has hardened into a protective cover called an elytron.

In many cases we experience "harmful" bug populations as an unintended consequence of our own action. We may introduce a new animal into an environment that has no offsetting predator. We may use insecticide or herbicide to modify our environment, and inadvertently kill off "beneficial" insects and enable previously benign insect populations to become "harmful." In the short term, an abundance of prey may cause the population of predators to swell, which leads to overconsumption of prey and a collapse of the predator population. Over the long term, an equilibrium exists in nature. Predators and prey are in balance if we leave them alone.

There are exception to "let the bugs control themselves."

- Invasive species that have no natural predator can multiply rapidly and cause extensive damage. The Emerald Ash Borer (*Agrilus planipennis)*, for instance, is an introduced beetle that is devastating many of our eastern forests. Left alone, its population will keep growing until all the host trees are gone.
- Urban stress (drought, pollution, soil compaction, heat…) can weaken trees and make them susceptible to various insects, including wood-boring beetles. These trees may need help to survive.
- Agricultural practices might lead to irruptions of beetles if host plants are continually grown in the same area. Some commercial farms might require pesticides to remain profitable.

Organization

For thousands of years people have been studying and classifying bugs. The best known system of classification (taxonomy) is the binomial nomenclature devised by Carl Linnaeus in the mid-1700s. While Linnaeus' system describes three Kingdoms of nature (Plant, Animal, and Mineral), we will only be concerned with animals.

Within the animal Kingdom, individuals are grouped into

- Phyla[5] (singular "Phylum") which contain
    - Orders (singular "Order") which contain
        - Families (singular "Family") which contain
            - Genera (singular "Genus") which contain
                - Species (singular "Species").

"Binomial nomenclature" refers to the Genus and Species of an individual. A Virginia Bigheaded Tiger Beetle for instance is in the Animal kingdom, Phylum *Arthropoda*, Class *Insecta*, Order *Coleoptera*, Family *Carabidae*, Genus *Tetracha*, and Species *virginica,* so its binomial nomenclature is *Tetracha virginica.* When written, it is generally referred to as *"Tetracha virginica"* when it first appears but is subsequently noted simply as "*T. virginica*." I have occasionally used that convention in this book. When the species of a bug is in doubt, "sp." will serve as a placeholder for "species."

While Linnaeus got things started, there are now more than one million organisms that have been classified. To prevent overcrowding, taxonomists have created numerous subdivisions such as "subphyla," "superorder," "suborder", "subfamily," etc., but the basic structure remains intact. The system has proven to be flexible enough to accommodate renaming and shifting of individuals within the system as more is understood about the true nature and relationships of the individuals. Increased use of genetic sequencing has led to frequent reclassification of insects and spiders. Classifications used in the volume are current and generally accepted as of 2021 but are all subject to revision as scientists learn more.

This volume has been organized to keep similar bugs together. Major categories correspond to taxonomic Orders: Within these major groups, beetles are grouped in taxonomic Families and, where appropriate, are further divided into Subfamilies and Tribes. Finally, they are listed alphabetically by genus and species. The index has a list of bugs by scientific name and by common name where those are available.

---

[5] All the bugs in this book belong in the Phylum Arthropoda, Order Coleoptera

Which Came First – The Beetle or the Egg?

Who hasn't pondered that question? When we think of the beetle (or the chicken) we generally think of the adult form. In the case of the chicken-egg controversy, we have the advantage of being familiar with both forms. We can envision both the egg (from the farm-fresh dozen in our refrigerator) and the chicken (smiling down from the sign above our local fried chicken diner.) We can unhesitatingly posit that the chicken is simply the egg's way of reproducing itself. With beetles, we are at a disadvantage. Few people have knowingly seen a beetle egg...and that's not the whole story. Beetles are holometabolous.

Holometabolism: It Isn't a Life Choice

For beetles, it's a long road from egg to adult. All beetles, as well as butterflies, moths, true flies, and Dobsonflies, are holometabolous. That is, they go through a complete four-stage metamorphosis: Egg, Larva, Pupa, and Imago (adult.)[6]

Eggs are generally laid in clusters (sometimes singly) on their larval food source and hatch quickly, often in a few days. This is necessary as they are subject to predation by wasps and flies and may be eaten by larval or adult beetles. There are exceptions. Some beetle species lay their eggs in protected areas (e.g., under tree bark) where they overwinter in a state of diapause. While most beetles abandon their clutch of eggs, carrion beetles are good parents, with both male and female guarding their eggs and feeding the larvae.

Larvae are often grub-like or worm-like, but there is great variety. One thing they have in common is that they are all engineered to eat. Eyes may be simple, but their mouth parts and digestive system are first rate. Japanese beetle grubs are perhaps best known beetle larvae, but other scarab beetle larvae look about the same and are just assumed to be

---

[6] Most other insects are hemimetabolous, going through a three-stage lifecycle of egg, nymph, and imago (adult.) The nymph frequently looks much like the adult, but without developed wings or reproductive parts.

Japanese beetles. While some larvae will retain the same form as they grow, others will transform significantly as they molt, with succeeding instars adopting different form and function. In any case, they will eat and molt several times before they are ready to pupate. At that point the

Scarab beetle larva

Click beetle larva

Lady beetle larva

larva will find a protected area (often underground) or will form a defensive surface within which it can transform. Some will enclose themselves in a wooden or dirt cell. Some will cover themselves with excrement for protection before pupating within their exoskeleton.

Pupae are generally stationary or have little capacity for movement while they transform. The beetles digestive juices start breaking down the animal itself. The dissolved material is then directed by special cells (histoblasts) to form the parts of the adult beetle. The entire process may take days or months, depending on species, nutrition, and environment.

Imagines (Adult beetles) break out of the pupal shell often looking soft and pale. At this point they are vulnerable to predation, but within a day or two they darken and their exoskeletons harden. Adults have fully developed wings (although a few beetles are flightless) and reproductive structures. Some beetles never eat in their adult form, so do not bother to develop a digestive system or robust mouth parts. Given the sheer number of species and their presence in all environments, the adults demonstrate behaviors and forms in dizzying variety. The primary function of all adult beetles is to mate.

How long does all this take? Some beetles have a life cycle that lasts only a couple months: Less than a week as an egg, six or seven weeks as a larva, a short period as a pupa, and a week as an adult. Others have long life cycles, most of which is spent in the larval or pupal stage. Larvae of hide beetles have been shown to regress to a less mature stage when faced with adverse conditions, then continue their development when conditions improve, sometimes taking years to mature. Sometimes beetles may enter diapause (the insect equivalent of hibernation) as larvae or pupae, extending their life cycle for decades.

With that, let's get started!

Size matters. Most insects are less than 1" long so they are generally measured in millimeters (mm.) and I use that convention in this book. There are 26 mm. in an inch. Images in this book are not to scale.

Speaking of size, when size is given in this book it refers to the body length, not including legs.

With thousands of beetles to choose from there is a high likelihood you will find bugs not mentioned in this book. At the end of the book, I have listed several excellent references to help you identify those.

Looking At Beetles

Our first look at a beetle is usually from above. They are small and generally cylindrical, a simple two-or-three-part beast with a half dozen legs and a pair of antennae. To understand how they work we need to take a closer look.

The dorsal (top) view of a beetle is relatively simple: A head, pronotum, and elytra. The elytra (hardened, modified forewings) can be raised to reveal the wings, which may lie flat or folded. There will be antennae which are sometimes tiny, sometimes longer than the beetle's body.

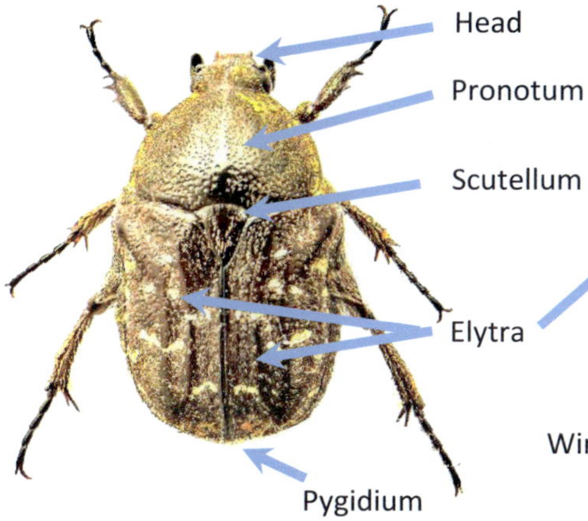

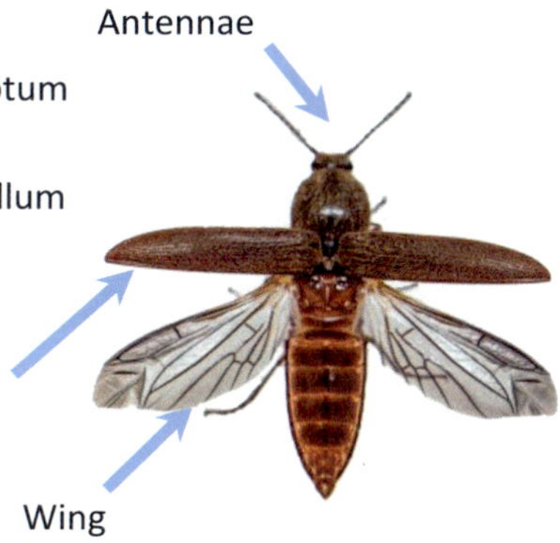

The underside has a few more moving parts. Hardened sclarites are connected by sutures that allow movement of the abdomen.

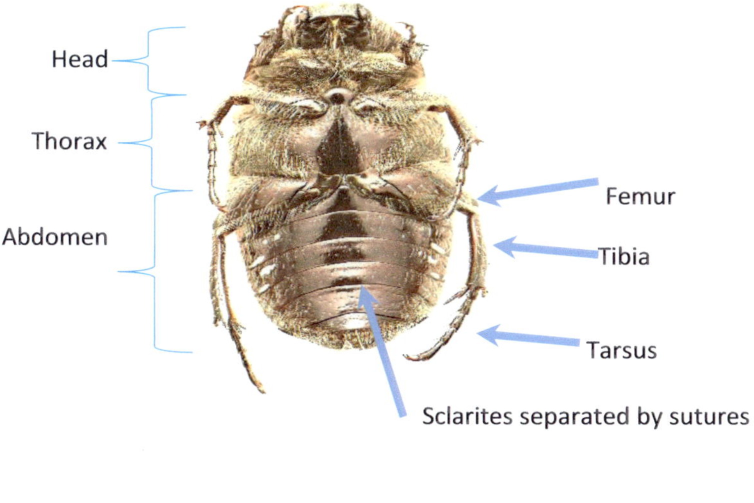

Order Coleoptera – The Beetles

With beetles we've hit the jackpot! There are about 400,000 species of beetles described and no doubt many more waiting in the shadows. Beetles constitute 40% of all described insect species. For each mammal species there are 100 beetle species. Chances are there are many different beetles in your yard, most of which you have never noticed.

Family Aderidae – Ant-like Leaf Beetles

As the name suggests, these small beetles vaguely resemble ants; often found on the underside of leaves.

Elonus hesperus
3 mm.

Family Anthicidae – Ant-like Flower Beetles

Cloudy Flower Beetle
(Anthicus cervinus)
3 mm.

Family Anthribidae – Fungus Weevils

Small weevils with straight, occasionally long antennae. Diet includes fungi and decaying plant material as adults, dead wood as immature.

Euparius sp (E. paganus?)
5 mm.

Family Brentidae – Primitive Weevils

Oak Timberworm
(Arrenodes minutus)
16 mm.

The Oak Timberworm (*Arrenodes minutus*) is the only Primitive Weevil found north of Florida. The male has a short, broad snout; the female has a longer, slender snout. As the name suggests, the preferred host for this weevil is oak where it drills into the tree to lay eggs.

Family Buprestidae – Metallic Wood-boring Beetles
(aka Jewel beetles)

15,000+ species, 78 reported in the U.S. They are, as the name suggests, a shiny and colorful group. Adult *buprestis* lives in dead or dying hardwoods, specializing in maple, beech, blackgum, oak, and elm. Larval *buprestis* burrow under the bark of trees and may live there for years before becoming adults. Normal populations of *buprestis* will not substantially weaken the tree, but if the tree is being grown to be harvested for lumber, the burrows made by larvae will devalue the wood.

Red-legged Buprestis
(Buprestis rufipes)
20 mm.

Dicerca obscura
13-20 mm.
Larvae found in common persimmon and sumac; adults on hickory, oak.

Family Cantharidae – Soldier Beetles

Aka "Leatherwings", soldier beetles are small, slender, and numerous (more than 500 species in the U.S.) Adults eat nectar, pollen, and other insects; larvae are fluid-feeding predators, feeding on insect eggs and larvae.

Two-lined Leather-wing
(Atalantycha bilineata)
8 mm.
aka Two-lined Cantharid

Wrinkled Soldier Beetle
(*Podabrus rugosulus*)
7-9 mm.

Margined Leatherwing
(*Chauliognathus marginatus*)
10 mm.

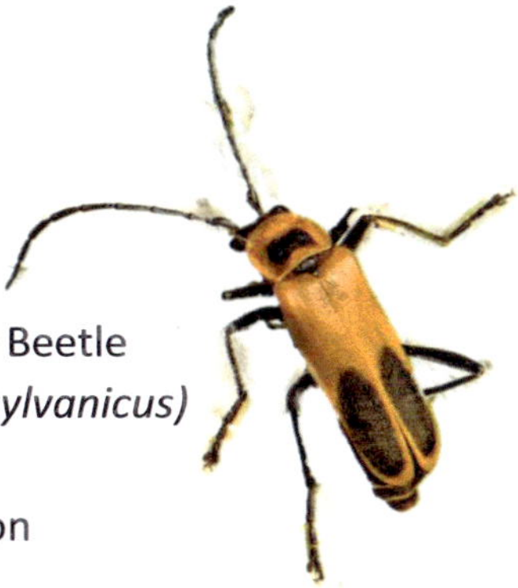

Goldenrod Soldier Beetle
(Chauliognathus pensylvanicus)
16 mm.
Quite common

Family Carabidae – Ground Beetles

Most of the 2,000+ species of Ground Beetles in the U.S. are shiny black or metallic with ridged wing-covers (elytra). Most are capable of producing defensive foul-smelling acidic secretions; some species are capable of shooting these chemicals when threatened. Generally considered beneficial, as they prey of the larvae of other garden insects.

Subfamily Carabinae

Fiery Searcher
(*Calosoma scrutator*)
35 mm.

Impressive in size and color, *C. scrutator* is uncommon and unforgettable. Observed only once in 2020. Also called "Caterpillar hunter," *C. scrutator* has an appetite for tent caterpillars and gypsy moths but will eat a variety of caterpillars and larvae. It is a top predator of the insect world.

Subfamily Cicindelinae (Tiger Beetles)

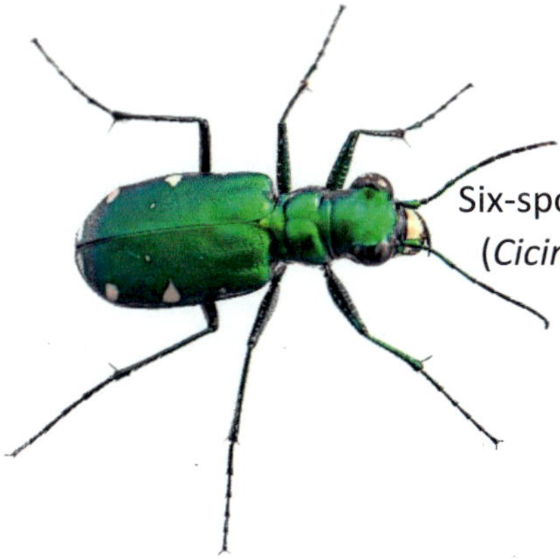

Six-spotted Tiger Beetle
(*Cicindela sexguttata*)
10-14 mm.

Virginia Bigheaded Tiger Beetle
(*Tetracha virginica*)
16-25 mm.

Punctured Tiger Beetle
(*Cicindelidia punctulata*)
10-13 mm.

Subfamily Harpalinae

Tribe Chlaeniini – (Vivid Metalic Ground Beetles)

Chlaenius aestivus
14-17 mm.
A Vivid Metallic Ground Beetle

Chlaenius laticollis
14 mm.

Chlaenius sp. nymph

Tribe Harpalini

Harpalus sp.
23 mm.

LeConte's Seedcorn Beetle
(Stenolophus lecontei)
6 mm.

Selenophorus ochropezus
5 mm.

Tribe Lebiini

Dromius piceus
7 mm

Lebia analis
5 mm.

Lebia grandis
10 mm.

Lebia viridis
6 mm.

Tribe Odacanthini

Cosnania pensylvanica
7 mm. ant mimic

Tribe Pterostichini (Woodland Ground Beetles)

Poecillus chalcites
10 mm.
A woodland ground beetle

Pterostichus sp.
More than 1,000 species in the genus. Most are predatory, though some feast on strawberries.

Subfamily Scaritinae

Pedunculate Ground Beetle
(*Clivina sp.*)
10 mm.

Family Cerambycidae – Long-horn Beetles

This large group of beetles is well represented in our area. Most longhorn beetles have quite long antennae, often longer than their bodies. Most *cerambicidae* larvae feed on woody tissue, while their adults are nectar eaters.

Subfamily Cerambycinae

Tribe Clytini

Red-headed Ash Borer
(Neoclytus acuminatus)
10 mm.

Locust Borer
(*Megacyllene robiniae*)
20 mm.

Tribe Elaphidiini

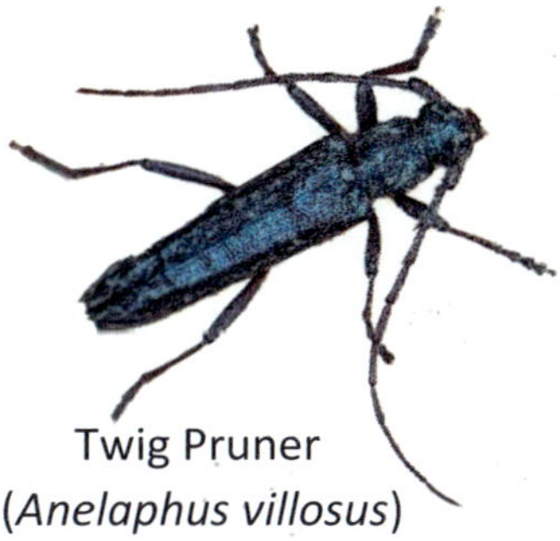

Twig Pruner
(*Anelaphus villosus*)
14 mm.

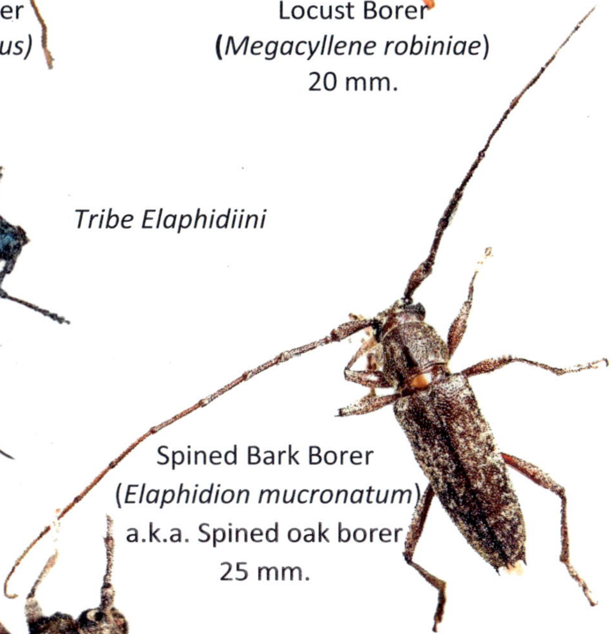

Spined Bark Borer
(*Elaphidion mucronatum*)
a.k.a. Spined oak borer
25 mm.

Enaphalodes atomarius
an oak borer
22 mm.

. Tribe Hesperophanini

Ash and Privet Borer
(Tylonotus bimaculatus)
9 mm.

Subfamily Lamiinae (Flat-faced Longhorns)

Tribe Acanthocinini

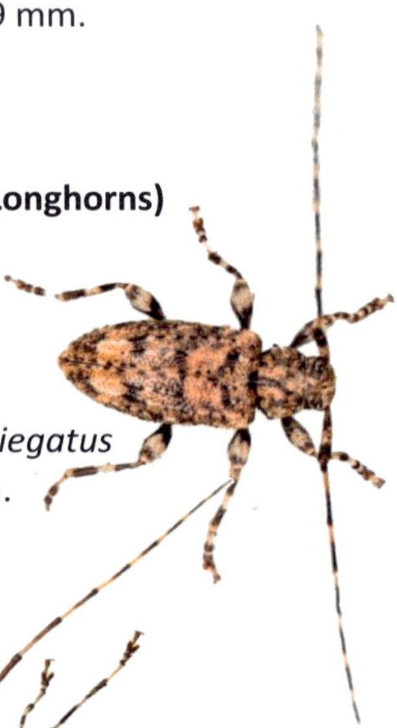

Astyleiopus variegatus
7-11 mm.

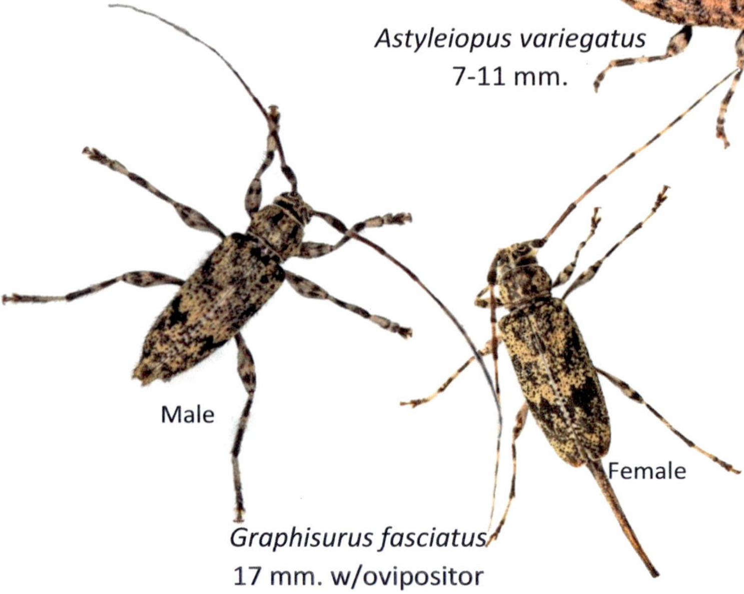

Male

Female

Graphisurus fasciatus
17 mm. w/ovipositor
Feeds on hardwoods and pine

Tribe Agapanthiini

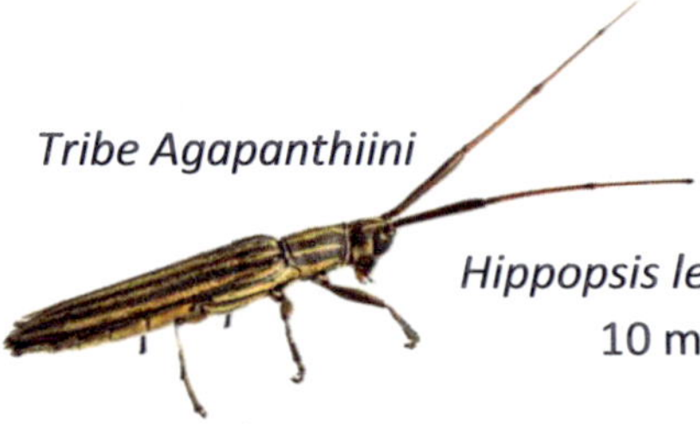

Hippopsis lemniscate
10 mm.

Tribe Monochamini

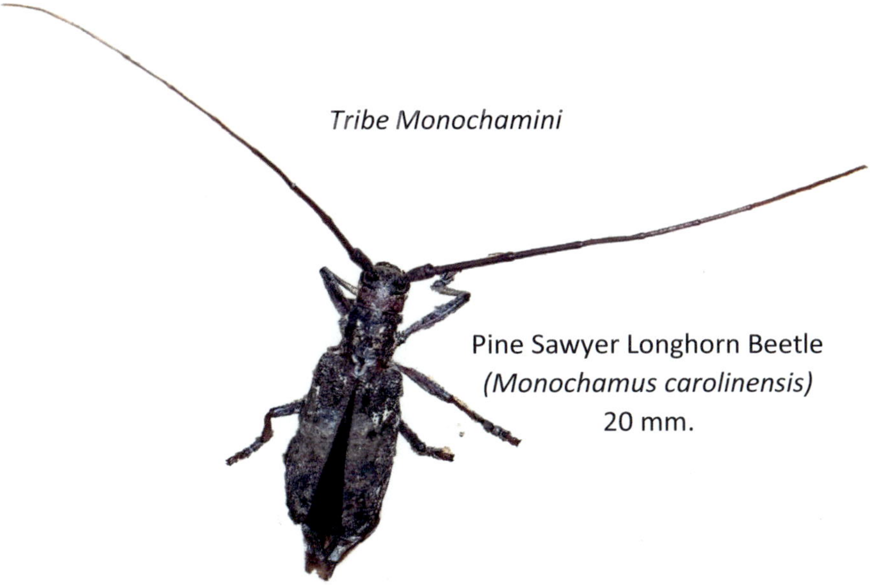

Pine Sawyer Longhorn Beetle
(Monochamus carolinensis)
20 mm.

Tribe Tetraopini

Red Milkweed Beetle
(*Tetraopes tetrophthalmus)*
12 mm.

Subfamily Lepturinae (Flower Longhorns)

Tribe Rhagiini

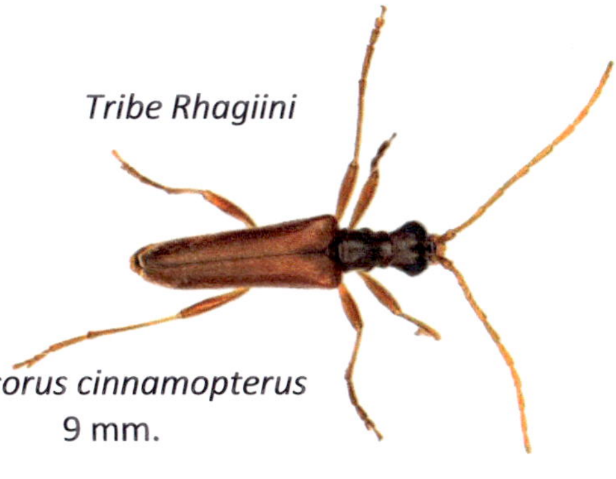

Stenocorus cinnamopterus
9 mm.

Tribe Lepturini

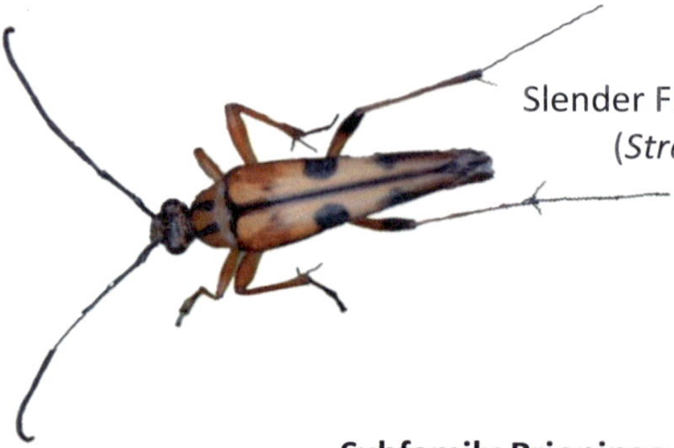

Slender Flower Longhorn Beetle
(*Strangalia famelica*)
9 mm.

Subfamily Prioninae

Tribe Prionini

Brown Prionid Beetle
(*Orthosoma brunneum*)
43 mm.

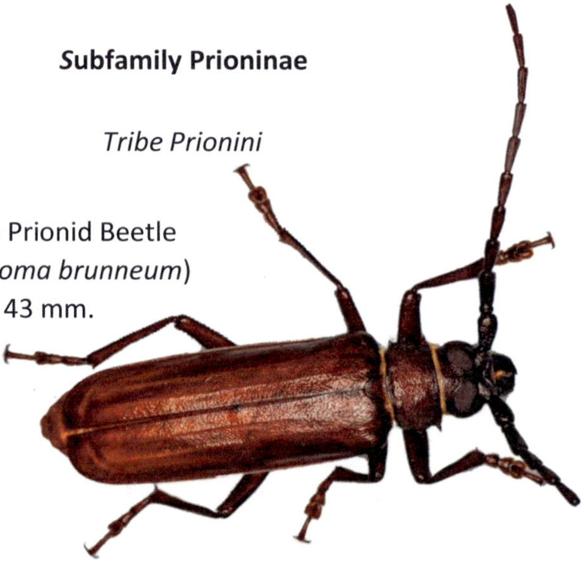

Tile-horned Prionus male
(Prionus Imbricornis)
40 mm.
Larvae feed in roots (oak, chestnut)

Family Chrysomelidae – Leaf Beetles

Generally small, often quite colorful, the leaf beetles munch on herbaceous plants. Several agricultural pests reside in this family.

Subfamily Bruchinae (Pea and Bean Weevils)

Acanthoscelides obtectus – a Bean Weevil
4 mm

Native to South America but now widespread, *A. obtectus* can be a major pest of beans and related legumes in the field and in storage. Yes, a bean weevil is a Chrysomelid, not a Curculionid like the rest of the weevils.

Subfamily Cassidinae (Tortoise Beetles and the Hispines)

Tribe Prionini

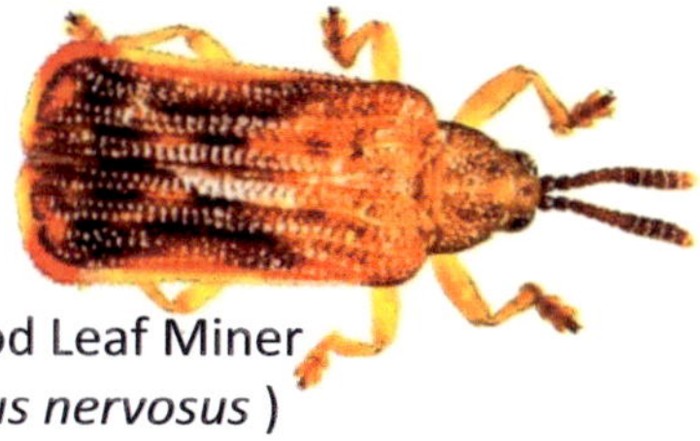

Basswood Leaf Miner
(*Baliosus nervosus*)
6 mm.

Tribe Cassidini (Tortoise Beetles)

The Mottled Tortoise Beetle (*Deloyala guttata*) nymph caries a fecal shield for protection. The adult sports a translucent shield of its own.
6 mm.

Subfamily Chrysomelinae

Swamp Milkweed Leaf Beetle
(*Labidomera clivicollis*)
12 mm.
A large, conspicuous beetle specializing in milkweed.

Subfamily Cryptocephalinae (Case-bearing Leaf Beetles)

Tribe Cryptocephalini

Cryptocephalus badius
4 mm.

Tribe Fulcidacini (Warty Leaf Beetles)

Neochlamisus bebbianae
4 mm.

Subfamily Eumolpinae

Tribe Adoxini

Demotina modesta
2 mm.
Definitely a female. D. modesta is parthenogenetic —there have been no males found in the US. Often found on oak.

Tribe Eumolpini

Brachypnoea sp
3 mm.
Found on a 'shroom

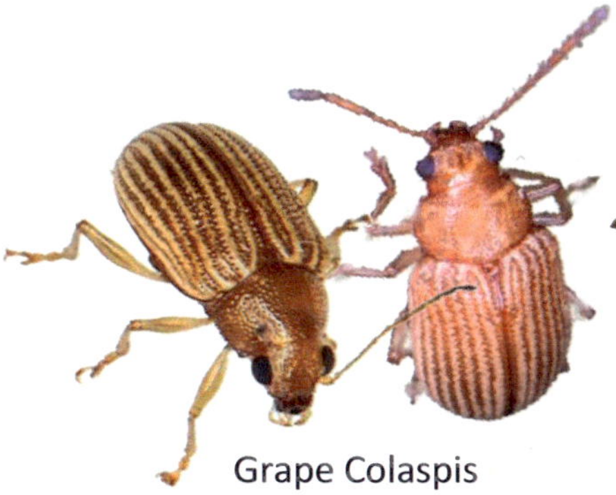

Grape Colaspis
(*Colaspis brunnea*)
4 mm.

Dogbane Leaf Beetle
(*Chrysochus aurantus*)
8-13 mm.
This all-green beetle reflects light in

Tribe Eumolpini (cont.)

Rhabdopterus praetextus
5 mm.

Note: Species identification requires inspection of the male genitalia; females cannot be identified except by association with the males. None of that was done. This bug is likely R. praetextus based on known distribution of this species.

Subfamily Galerucinae (Skeletonizing Leaf Beetles and Flea Beetles)

Tribe Alticini (Flea Beetles)

Charlie Brown Flea Beetle
(*Capraita sexmaculata)*
3 mm.

A Flea Beetle
(Crepidodera sp)
3 mm.

Tribe Galerucini

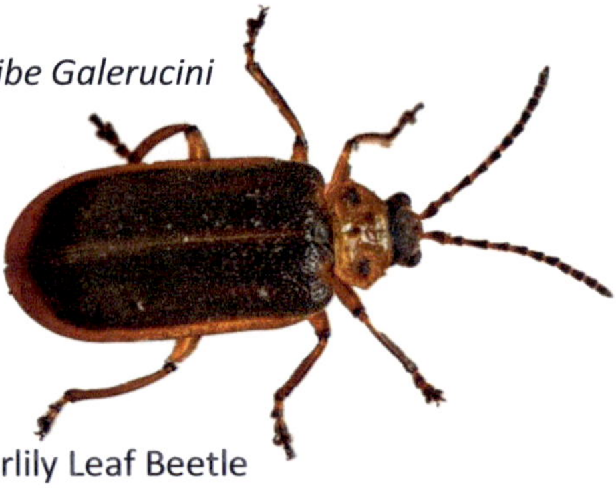

Waterlily Leaf Beetle
(*Galerucella nymphaeae*)
5 mm.

Many of these found on water lily leaf pads. Eggs are laid on top of the lily pads in clusters.

Tribe Luperini

Striped Cucumber Beetle
(*Acalymma vittatum*)
5 mm.

A major pest of back yard gardens, the adults will eat leaves, flowers and rinds of cucumbers and related crops; the larva will tunnel into the stems and vines. These will show up in large numbers even if you are planting your first cucumbers and there are no other plants in the neighborhood!

Bean Leaf Beetle
(*Cerotoma trifurcate*)
5 mm.

Cucumber Leaf Beetle
5 mm.
(*Diabrotica undecimpunctata)*

Aka Spotted Cucumber Beetle or Southern Corn Rootworm, this bug can cause considerable damage to your crops. The larvae eat the roots of young plants, while the adults eat the flowers, foliage, and fruit.

Family Cleridae – Checkered Beetles

Narrow and hairy, checkered beetles fill many ecological niches. Some are used as biological controls, as they have a healthy appetite for bark beetles: adults eating adults, larvae eating larvae.

Subfamily Clerinae

Priocera castanea
6 mm.

Subfamily Epiphloeinae

Madoniella dislocata
3-6 mm.

Family Coccinellidae – Lady Beetles

Known as ladybugs in the U.S. and ladybird beetles in the rest of the English-speaking world, lady beetles are distinctive, beneficial, and beloved. Most are consumers of aphids and scale insects, although some are predators of other lady beetle nymphs.

Seven-spotted Lady Beetle
(*Coccinella septempunctata*)
7 mm.

Spotted Lady Beetle
(*Coleimegilla maculata lengi*)
5 mm.

Polished Lady Beetle
(*Cycloneda munda)*
4 mm.

Asian Lady Beetle
(*Harmonia axyridis*)
6-8 mm.

Imported for pest management, *H. axyridis* is larger and more aggressive than most native lady beetles. The larvae and adults eat the larvae of native lady beetles as well as aphids. As a result, *H. axyridis* is now the dominant species in most of its range. This species is quite variable in pattern.

Twenty-spotted Lady Beetle larva
(Psyllobora vigintimaculata)

Tribe Scymnini (Dusky Lady Beetles)

Scymnus sp

Not all lady beetles are red.
Some are dark and hairy.
3 mm.

Families Cucujidae and Silvanidae – Flat Bark Beetles

Family Cucujudae - Flat Bark Beetles

Red Bark Beetle
(Cucujus clavipes)
12 mm.

Family Silvanidae - Silvanid Flat Bark Beetles

Telephanus atricapillus
4 mm.

Family Curculionidae – Snout[7], Ambrosia, and Bark Beetles

Subfamily Baridinae - Flower Weevils

Corn Pith Weevil
(Geraeus penicillus)
2 mm.

Subfamily Condamine

Cylindrocopturus quercus
2 mm.

Subfamily Curculioninae

Yellow Poplar Weevil
(*Odontopus calceatus)*
4 mm

A.k.a. sassafras mining weevil, tulip tree leaf miner, tulip tree weevil, or the magnolia leaf miner. Adults emerge in the spring after overwintering in the ground. They lay eggs on the leaves of host plants (yellow poplar, magnolia...). By mid-summer the eggs have hatched, the larvae have pupated, and the adults disappear into the leaf litter till next year. Although not usually harmful to their host, occasional infestations have caused significant damage to yellow poplar.

---

[7] Weevils

Subfamily Cyclominae

Listronotus sp.
looks like *L .appendiculatus*
6 mm
This small weevil had even smaller beetles (*Toramus pulchellus)* riding on its back.

Subfamily Dryophthorinae

Ironweed Curculio
(*Rhodobaenus tredecimpunctatus)*
10 mm.

Sphenophorus melanocephalus
A 10 mm. billbug
Grubs will eat the roots of turf grass. The grubs look like small Japanese beetle grubs but are straight rather than "C"-shaped; they have no legs.

Maize Weevil
(*Sitophilus zeamais)*
3 mm.

Fourteen species of the Sitophilus (granary weevil) genus are found throughout the world feasting on grains and nuts, causing considerable loss of stored grains. This individual showed up in a sack of feed corn, so is probably *Sitophilus zeamais*, the maize weevil.

Subfamily Entiminae (Broad-nosed Weevils)

Tribe Cyphicerini (Oriental Broad-nosed Weevils)

An Asian import, *Pseudoedophrys hilleri* tends to prefer maple and witch hazel. That is certainly the case of these particular individuals who are abundant under a maple tree, although they have not caused noticable damage to the tree.

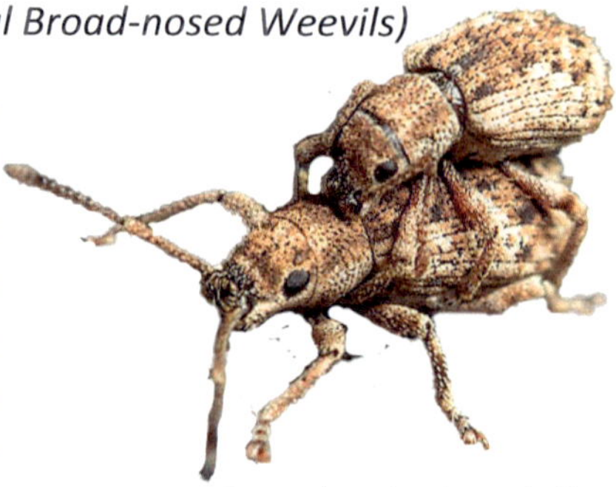

Pseudoedophrys hilleri
4 mm.

Tribe Naupactini

Pseudocneorhinus bifasciatus, the Two-banded Japanese Weevil, was introduced from Asia about 100 years ago and is now well established throughout the eastern U.S. With more than 100 host plants, this weevil seems to have a particular fondness for roses, hemlock and laurel. All *P. bifasciatus* in the U.S. are female, reproducing parthenogenetically

Two-banded Japanese Weevil
(Pseudocneorhinus bifasciatus)
3 mm.

Tribe Trachyphloeini

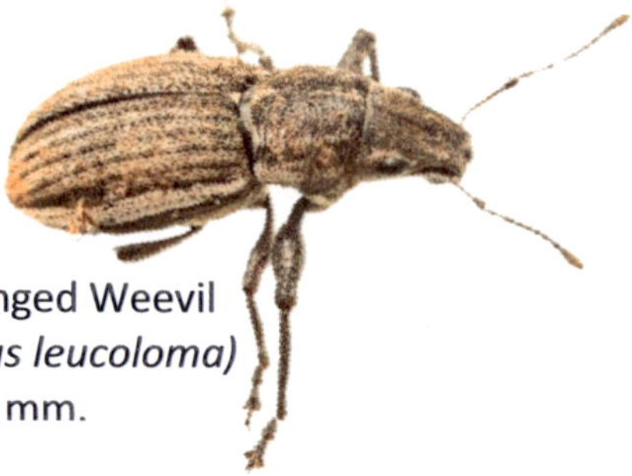

White-fringed Weevil
(*Naupactus leucoloma)*
10 mm.

Subfamily Lixinae

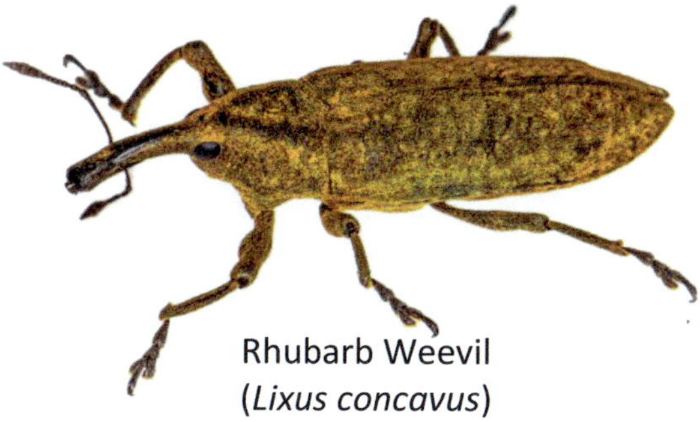

Rhubarb Weevil
(*Lixus concavus*)
12 mm

The Rhubarb weevil is known to cause damage to rhubarb by cutting into the stems although it is rare to find this species producing economically significant damage to crops. The larva may damage roots of crops.

Subfamily Molytinae

Pecan Gall Curculio
(*Conotrachelus elegans*)
4 mm.

Micralcinus sp.
M. parvulus or *M. stehi*
3 mm.

Subfamily Scolytinae (Bark and Ambrosia Beetles)

Euwallacea validus
2 mm.

An invasive ambrosia beetle from Japan, first noted in New York in 1975, is now broadly established in the United States. These beetles bore into the phloem layer of various trees, where they release spores and cultivate fungus gardens for food. This one was found on an old, weakened cherry tree. Looking unlike the snout beetles (weevils), ambrosia beetles were formerly classified as their own family, Platypodidae. They are now considered Scolytinae, a sub-family within Curculionidae.

Xyleborus xylographus
3 mm.

A bark beetle. Looking unlike the snout beetles (weevils), bark beetles were formerly classified as their own family, *Scolytidae*. Like *E. validus* above, they are now considered Scolytinae.

Family Dermestidae – Skin Beetles

Although adult skin beetles feed on plants, the larvae feed on dried skin, fur, fibers, and feathers. Some species are useful in forensic analysis.

Anthrenus sp
3 mm.

Anthrenus verbasci
3 mm.

Trogoderma sp.
a carpet beetle
3 mm.

Family Elateridae – Click Beetles

A large family whose shared characteristic is the ability to snap their bodies with a "click", propelling themselves into the air. While the adults have little agricultural impact, several species' larvae (wireworms) cause significant damage.

Subfamily Agrypninae

Tribe Hemirhipini

Eyed Click Beetle
(*Alaus oculatus*)
40 mm.
The largest of our click beetles

Tribe Oophorini

Flat Wireworm
(*Aeolus mellilus*)
6 mm.

Conoderus lividus
11-17 mm.

Conoderus bellus
5 mm.

Subfamily Dendrometrinae

Gambrinus griseus
10-14 mm.

Subfamily Elaterinae

Tribe Elaterini

Elater abruptus
15-22 mm.

Orthostethus infuscatus
30 mm.

Tribe Melanotini

Melanotus sp
5-12 mm.

Family Elmidae – Riffle Beetles

Larvae and adults are generally aquatic, living under rocks and eating algae.

Stenelmis sp.
7 mm.

Although most riffle beetles are aquatic, this individual found its way ashore (about 25' from the lake.)

Family Erotylidae – Pleasing Fungus Beetles

Beetle of little impact on human endeavors, this family goes about life eating fungi and plant material; some are pollinators.

Tribe Erotylinae

Four-spotted Fungus Beetle
(Ischyrus *quadripunctatus)*
9 mm.

Megalodacne fasciata
9-15 mm.

Tribe Xenoscelinae

Toramus pulchellus
1 mm.

Family Mycetophagidae – Hairy Fungus Beetles

Another family of fungus eaters, the Myvetphagidae live quietly among the leaf litter and fungi.

Litargus tetraspilotus
2 mm.

Family Gyrinidae – Whirligig Beetles

The most visible of the aquatic beetles, the whirligigs are easily seen in groups swimming on the surface of the water. When disturbed they swim in a frenzied circular pattern, hence the "whirligig." The eyes are split, allowing sight simultaneously above and below water.

Dineutus sp
11 mm.

Family Haliplidae – Crawling Water Beetles

This family of water beetles does not swim well, preferring to crawl. Their swimming is accomplished by paddling alternately, creating a clumsy-appearing motion. This may be beneficial as a means of confusing predators.

Peltodytes muticus
5 mm.

Family Heteroceridae – Variegated Mud-loving Beetles

Small beetles that dig (sometimes tunnel) through mud at the edge of ponds, eating algae.

Heterocerus mollinus
5 mm.

Family Histeridae – Clown Beetles

Predacious beetles with short elytra (wing covers), clown beetles will eat larvae, munch on carcasses, and eat each other.

Subfamily Abraeinae

Aeletes politus
1 mm.

Subfamily Saprininae

Euspilotus assimilis
5 mm.
A common carrion-feeder

Family Hybosoridae – Scavenger Scarab Beetles

A Pill Scarab Beetle
(*Germarostes globosus*)
3 mm.
Nearly identical to *Germarostes aphodioides*

Family Hydrophilidae – Water Scavenger Beetles

The water scavenger beetles are misnamed, as the larva are predators and, while some adults will scavenge, most will predate as well. Mosquito larvae are a favorite of some *Hydrophilidae*.

Subfamily Hydrophilinae

Collared Water Scavenger Beetle
(*Tropisternus collaris*)
8 mm.

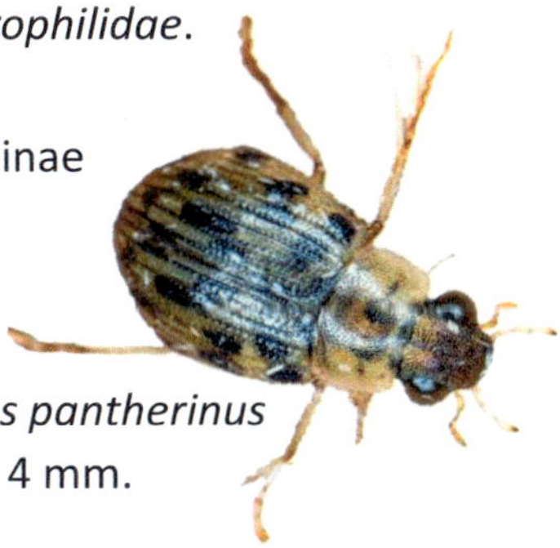

Berosus pantherinus
4 mm.

Subfamily Sphaeridiinae

Cercyon praetextatus
8 mm.

Family Lucanidae – Stag Beetles

Large, distinctive beetles. The males have oversized mandibles reminiscent of the antlers of a stag. Larvae of the stag beetles may live several years and several larval stages before pupating. The males may look intimidating, but they are not aggressive toward humans. The smaller females have much smaller (and more powerful) mandibles.

Yellow-thighed Stag Beetle male
(*Lucanus capreolus*)
a.k.a. Reddish-brown Stag Beetle
28 mm.

Family Lycidae (Net-winged Beetles)

End Band Net-winged Beetle
(Calopteron terminale)
12 mm.

Family Meloidae – Blister Beetles

Adult blister beetles are all capable of producing cantharidin, a blister agent, for defense. These insects may inhabit grasses—especially alfalfa- - which could be processed into hay bales. A few beetles consumed in a single feeding may be lethal to a horse.

Margined Blister Beetle
(*Epocauta funebris*)
10 mm.
Pest of tomatoes or potatoes.

Black Blister Beetle
(*Epocauta pensylvanica*)
11 mm.

Striped Blister Beetle
(*Epocauta vittata*)
9 mm.

The larvae remain mostly underground, feeding on grasshopper eggs. The adults, however, are a major pest of squash, beans, and numerous other crops. They tend to appear in swarms. If you pick these off your plants, wear gloves. They are blister beetles, after all.

Family Melyridae – Soft-winged Flower Beetles

Primarily pollen-eaters as adults, the Melyridae are sometimes called "pollen beetles."

Attalus sp
6 mm.

Family Mordellidae – Tumbling Flower Beetles

To escape predators the tumbling flower beetle will initiate a series of seemingly uncoordinated jumps, appearing to tumble. Due to the long abdominal tip that facilitates these movements, they are sometimes called pintail beetles.

Falsomordellistena pubescena
4 mm.

Family Nitidulidae – Sap-feeding Beetles

Small beetles that feed on sap or over-ripe fruit.

Subfamily Cryptarchinae

Attracted to beer, wine, fruity or fermenting beverages, the picnic beetle earns its name by drinking from and drowning in your beer.

Picnic Beetle
(*Glischrochilus fasciatus*)
5 mm.

Glischrochilus sanguinolentus
5 mm.

Glischrochilus siepmanni
6 mm.

Four-spotted Sap Beetle
(*Glischrochilus quadrisignatus*)
5 mm

Associated with milkweed seed pods, this insect is sometimes referred to as "beer bug" or "picnic beetle" for its habit of finding and drowning in your beer.

Subfamily Epuraeinae

Epuraea sp
3 mm.

Subfamily Nitidulinae

Strawberry Sap Beetle
(*Stelidota geminata*)
3 mm.
Quickly attracted to damaged or over-ripe strawberries, *S. geminate* can become a pest in the garden.

Family Ptiliidae (Featherwing[ed] Beetles)

2 mm.
A consumer of fungi and mold.

Family Ptilodactylidae – Ptilodactylid Beetles

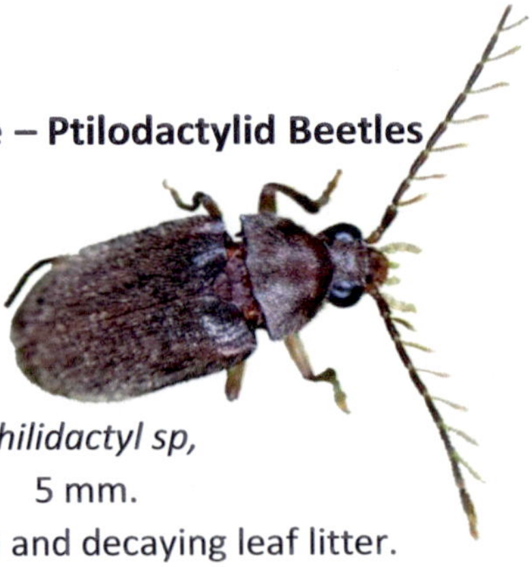

Philidactyl sp,
5 mm.
A consumer of fungi and decaying leaf litter.

Family Ptinidae – Death Watch & Spider Beetles

Caenocara sp, a puff-ball beetle
2 mm.
Small beetle that develops within puffball fungi.

Family Pyrochroidae – Fire-colored Beetles

Also known as "eyelash bugs," these are primarily fungus-eaters

Neopyrochroa femoralis
13 mm.

Neopyrochroa flabellate
13 mm.
A nocturnal fungivore, spends days under decaying logs.

Family Scarabaeidae – Scarab Beetles

A large and important family, scarabs fill many niches. Some are recyclers of dung (dung beetles) while others are plant eaters (chafers, Japanese beetles.) The scarab revered in ancient Egypt is a dung beetle.

Subfamily Aphodiinae (Aphodiine Dung Beetles)

Black Turfgrass Ataenius
(Ataenius spretulus)
4 mm.

Calamosternus granaries
5 mm.

Subfamily Cetoniinae (Fruit and Flower Chafers)

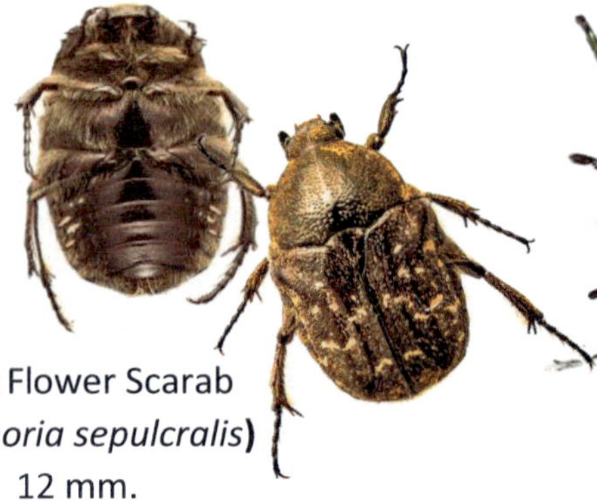

Dark Flower Scarab
(*Euphoria sepulcralis*)
12 mm.

Green June Beetle
(Cotinis nitida)
15-27 mm.

Delta Flower Scarab
(*Trigonopeltastes delta*)
10 mm.

Subfamily Dynastinae (Rhinoceros Beetles)

Northern Masked Chafer
(Cyclocephala borealis)
14 mm.

Eastern Hercules Beetle (F)
(*Dynastes tityus*)
52 mm.

Subfamily Melolonthinae (May Beetles and Junebugs)

Rose Chafer
(*Macrodactylus subspinosus*)
7-10 mm.

Asiatic Garden Beetle
(*Maladera formosae*)
12 mm.

Subfamily Rutelinae (Shining Leaf Chafers)

Oriental beetle
(*Exomala orientalis*)
7-13 mm.

Japanese Beetle
(*Popillia japonica*)
11 mm.

Few beetles are as common, numerous, and destructive as the Japanese beetle. The adult has an appetite for more than 300 plants, and its grubs are a major destroyer of turf.

Grapevine Beetle
(*Pelidnota punctata*)
24 mm.

Subfamily Scarabaeinae (Dung Beetles)

Scooped Scarab Beetle
(*Onthophagus hecate*)
A dung beetle

Family Silphidae – Carrion Beetles

Where would we be without carrion beetles? Knee-deep in carcasses? Probably not. There are a host of other recyclers at work. However, carrion beetles are some of the first, fastest, and most efficient recyclers of animal bodies. An additional benefit: These beetles eat many of the fly larvae found on carrion. This family is also noted for parenting behavior, protecting and feeding their larvae.

Red-lined Carrion Beetle
(Necrodes surinamensis)
12-24 mm.
Adults and nymphs feed on carrion and fly larvae.

American Carrion Beetle
(*Necrophila americana*)
20 mm.

nymph

While flies are often the first insects at a corpse, *N. americana* arrives soon after and provides an important service: Not only does it eat the carrion, it also eats the fly larvae. The beetle lays and guards its eggs; the nymphs are active after other recyclers have departed, so consume more dried hide, hair, or feathers. One generation per year.

Gold-necked Carrion Beetle
(*Necrophila tomentosus*)
24 mm.

Ridged Carrion Beetle
(*Oiceoptoma inaequale*)
14 mm.

Margined Carrion Beetle
(*Oiceoptoma noveboracense*)
12 mm.

Family Staphylinidae – Rove Beetles

With 63,000 members, the rove beetles comprise the largest family of organisms. They are predators, generally long and thin with short elytra.

Subfamily Aleocharinae

Aleochara littoralis (?)
A parasitic rove beetle
6 mm.

Astenus sp

Falagria dissecta
2 mm.

Meronera venustula
3 mm.

Subfamily Oxytelinae (Spiny-legged Rove Beetles)

Apocellus sphaericollis
4 mm.
The only *Apocellus* species in North America

Subfamily Paederinae

Caution: Do not crush! Some Paederinae **contain pederin[8] which will cause severe blistering if it gets on your skin.**

Tribe *Lathrobiini*

Astenus sp.
3 mm.

Scopaeus sp

Rugilus sp.
3 mm.

Lathrobiina (sub-tribe) sp

[8] Pederin is manufactured not by the beetles, but by endosymbiont bacteria.

Tribe *Paederini*

Paederus littorarius
6 mm.

Caution: This is one of the species that contain significant levels of pederin. If it is crushed or dragged across your skin, blistering will start within 12-36 hours and effects may last several weeks. Best to wash your hands and any affected area if you come in contact with this beetle.

Homaeotarsus sp.
10 mm.

Tribe *Pinophilini*

Pinophilus sp.
probably *Pinophilus latipes*
14 mm.

Subfamily Staphylininae (Large Rove Beetles)

Belonuchus rufipennis
7 mm.

Subfamily Tachyporinae (Crab-like Rove Beetles)

Coproporus ventriculus
3 mm.

Tachyporus sp,
3 mm.

Family Tenebrionidae – Darkling Beetles

A large family of dark beetles, primarily scavengers of dead wood and debris but also know to dine on other insects. Larvae resemble click beetle larvae.

False Mealworm Beetle
(*Alobates pensylvanicus*)
21 mm.

Family Trogidae – Keratin (Hide) Beetles

Among the scavengers, Hide beetles are late arrivals at any carcass. They specialize in hide, hair, and feathers. Some may be found in bird nests. These bugs are not widely studied.

Trox unistriatus
11 mm.
Found on a fox carcass

References

Excellent reference material is available in print and on the Internet. These are some of my favorites:

Print

Evans, Arthur V., **Beetles of Eastern North America**, Princeton University Press, Princeton 2014. This is my go-to reference for beetles.

Eaton, Eric R. and Kaufman, Kenn, **Kaufman Field Guide to Insects of North America**, Houghton Mifflin, New York, 2006

Gullan, P.J. and Cranston, P.S., **The Insects – *An Outline of Entomology***, Wiley Blackwell, Chichester, 2014 – If you slept through your college entomology class, this is what you missed.

Walters, Martin, **The Illustrated World Encyclopedia of Insects**, Lorenz Books, London 2017

Marshall, Stephen A., **Beetles – The Natural History and Diversity of Coleoptera**, Firefly Books, Buffalo 2018. Huge, comprehensive, and informative, but not as accessible as Evans' *Beetles of Eastern North America* for our local beetles.

Internet

BugGuide (bugguide.net) –Post an image of your unknown bug and the entomologists, grad students and assorted experts who support BugGuide will identify it for you, often to the species level. Response is generally timely (within one or two days). If you already know the bug you can enter the name and be directed to pictures and information on the species.

INaturalist (inaturalist.org) Another place you can upload images and get community help with identification. Unlike BugGuide, INaturalist maintains an image recognition algorithm that will suggest a tentative ID pending human confirmation. A useful feature is graphic displays of when and where sightings were reported.

Ground Beetles – Tribe Harpalini (https://www.pbase.com/tmurray74/ground_beetles_tribe_harpalini) Lots of ground beetles look about alike. Photos on this site are not diagnostic but might help steer you toward an ID.

Virginia Beetles (https://www.beetleidentification.org/beetles-by-state-listing.php?reach=Virginia) Good visual guide to our local beetles

Virginia Insects (https://www.insectidentification.org/insects-by-state.php?thisState=Virginia) An excellent visual guide to more than 1,000 insects found in Virginia

Google (google.com) Google, of course, will find information and images of whatever bug interests you. Often Wikipedia is the first hit.

Index

Binomial Nomenclature

ABOUT THE AUTHOR

Educated as a chemist, Lee Alloway was an intern at NASA, a military pilot, and an advisor to the Hungarian Ministry of Defense before serving fourteen years as a civil servant in the Pentagon. He got his first camera when he was five and has documented most of his travels through more than 100 countries. His interest in macro photography led to a passion to study the insects and other arthropods that live unnoticed around us. For the past decade he has traveled around the world photographing bugs, often in the company of entomologists and other macro photographers. He is the author of eight books of poetry and a large collection of seldom heard songs. This is his third book dedicated to bugs of Northern Virginia.

Notes

Notes

Notes

Notes

Made in United States
Troutdale, OR
11/03/2026

56844870R00048